The Truth About DRUGS And ADDICTION

By

Justin John Carroll

Contents

Introduction

PART ONE: WHY WE TAKE DRUGS

1. Partners in Crime – Dopamine and Serotonin
2. We're All Addicts
3. Temptation
4. Behind Closed Doors
5. The Power of Association

PART TWO: RE-TRAIN YOUR BRAIN

6. Why The Drugs Won't Work
7. Withdrawal And Relapse
8. Rehabilitation – Healthy Resources
9. The 'Rebel' Myth

Conclusion

Bibliography

Introduction

I started smoking first, when I was 16, and started drinking shortly after that, in the same year. When I was 18, I started drinking coffee, and later, tea. I'd never had tea or coffee before that time.

During my time at University (where I failed to obtain a degree due to addiction), I tried a number of other drugs including LSD, Amphetamine, Cannabis, Valium, Diamorphine, Amyl Nitrate, Beta Blockers, Butane Gas, and Guarana. Later on in my twenties, I added Ecstasy, Magic Mushrooms, Cocaine and Pain Killers to the list. The taking of all these other drugs was directly encouraged by alcohol consumption. They say Cannabis is the 'gateway' drug. It's not. Alcohol is.

And the one drug that goes largely overlooked that starts it all by creating the Adult Addict blueprint when you're still just an innocent kid...that's right! It's Refined Sugar. Take a bow.

Lest we forget, you can also be addicted to compulsive, non-chemical (or less chemical) behaviours such as Sex, watching TV, watching Movies, Overeating, eating Junk Food, Masturbation, and even Exercise which, in an unhealthy attachment, can be just as mind-bending as a drug. People might call it an obsession, but it's not: it becomes an addiction. In fact, Exercise gives you the same high the drug addict seeks. They're just two different paths to the same end result.

PART ONE:

WHY WE TAKE DRUGS

Chapter One:

Partners in Crime – Dopamine and Serotonin

What's D+S? Thanks for asking. It stands for Dopamine and Serotonin. 'What have they got to do with addiction?!' I hear you cry. Well, everything. In fact, they ARE addiction.

In a massive oversimplification, Serotonin makes you feel good, and Dopamine is a central constituent of reward-motivated behaviour (which, sadly, begins in childhood with parents 'rewarding' their children for 'good' behaviour with sugary treats – the refined sugar provokes an unnecessary flood of Dopamine and Serotonin, and the brain quickly learns that this is the fastest way to obtain these biochemicals which means from that point on, the brain will seek out any other triggers that have the same effect. Thus, the blueprint for addiction – or unhealthy attachment to manufactured catalysts – is laid down and ingrained. Remember, this happens over a period of several years).

In some people's systems, the ability to produce regular, healthy or normal levels of D+S is malfunctioning – usually because of early trauma, or an unhealthy environment during childhood and teens. There are not enough receptors in the brain for the D+S to cling to, so they need to produce larger amounts to cling to each receptor to achieve the same effect as a person with a healthy, balanced system. Unfortunately, as this becomes the pattern, fewer receptors are needed and a capacity that was already depleted becomes even more depleted. Then even more D+S release is required, so even more 'triggers' are required: it is indeed a vicious cycle, and explains why all addictions become progressively more acute and intense.

The way the larger amounts of D+S are produced in the brain is by artificially stimulating that production. The easiest and quickest way to do this is by introducing a toxin, or poison, into the system. The body believes it is under attack and releases its own biochemical opiates, or Endorphins (endogenous MORPHINES) – Dopamine and Serotonin – in order to cope with the problem. That's why all drugs are TOXINS. Since the brain is such an efficient organism, once it has learned this, it will not want to return to slower ways (which are non-poisonous, healthy ways) of producing the D+S – ways which are infinitely more wholesome and healthy, like exercise. It's actually trying to do you a favour!

Dopamine is released naturally in the Mother-Child nurturing process, and I believe, stimulated by the vibrational frequency we call 'love'. So, effectively, you are attempting to nurture yourself in a loving way by taking drugs. Sounds fucked up, doesn't it? It is.

All drug-taking is really COMFORT-seeking behaviour. You're after comfort, relief or elation (a 'high') – but mainly COMFORT. In other words, you want to feel better. Ever heard Ecstasy addicts talk about how they 'just love everyone' when they're high? Sadly, all drug-taking is actually an attempt to feel loved, feel love for others, or feel love for yourself, and that's the essence of the comfort we seek.

The healthy way – the way we were actually designed – to feel loved and feel love for ourselves and others is to be high in spirit. Hence, spirit-uality. You, literally, raise your vibrational frequency from low to high, or if you like, from heavy to light, from negative to positive, from DARK to LIGHT. This is non-destructive and truly life-enhancing. Taking drugs to achieve all this is a sad and poor man-made replication of one of the greatest gifts to Humanity.

The perceived 'wild excitement' of taking drugs is actually an illusion. It is a thorough falsehood.

To get back to D+S....you might be wondering at this point why some people have fewer D+S receptors in the brain. Hang on to your Politically Correct Hats.

It may be simple biology. DNA. There may be some biochemical reason why a person would have fewer receptors than others. However, any research I've ever done has shown that an adequate number of receptors do not develop without the adequate stimulation to do so when the person is a child, and more especially, an infant. It would seem that 99% of it comes down to nurture and environmental influences, not nature. The adequate stimulation required is love – UNCONDITIONAL love – and especially the love shared in the Mother-Child bond. This is why Mothering is absolutely the most important 'job' on the planet. Mothers literally create the new generations. That's not to say a Father's Role isn't equally important in a different way. But a Mother's natural energy – assuming she is balanced and healthy – is geared towards loving, caring for, and nurturing the infant. Essential in rearing a healthy child. A Father brings support (of the Mother), love, guidance, discipline and protection.

These simple roles have become enormously confused in recent decades.

I would also posit the theory here that receptors in the brain, even if at a healthy level to begin with, can be reduced in childhood or during adolescence by any Mother/Father behaviours that are contrary to the ones I just described. For example: neglect, bullying, withdrawal of support, threats, physical or sexual abuse, parental relationship disharmony, divorce or separation, verbal abuse, dogmatism, distancing, lack of guidance, poor examples of social behaviour, prolific or unnecessary cursing, and the Big One...parental drug use as well as drug abuse. In drug use, I include Alcohol, Nicotine and Caffeine since most 'adults' partake of one of these, a combination of these, or all of these. And make no mistake – the parent's use of these drugs influences the child growing up by normalising their use, and in the child's mind, linking that use to being an adult!

Now I posit a second theory having read Dr. Masaru Emoto's 'Hidden Messages in Water' which clearly demonstrates the effect of negativity (eg. Swearing) on the formation of water crystals. The average adult body is 50% - 70% water, and the average child is about 75% water. More significantly, perhaps, the human brain is approximately 85% water. If something as simple as swearing can negatively affect the formation of water crystals, and we are largely made of water, what impact do you think harsher, more cruel behaviours (especially if repeated) might have on children's and adolescents' brains in particular?

Dr. Emoto's research also demonstrates that even just the words 'peace' and 'love' can encourage the water crystals to reform in the most exquisite, intricate and symmetrical patterns!

So looking for stimulants to produce more D+S is a kind of love-replacement therapy!

Chapter Two:

We're All Addicts

Drug addicts. Alcoholics. They're the bad guys, right? Wrong! In western 'civilization', we're all drug addicts (to a greater or lesser extent), even the children. Too much TV, too many computer games, too many refined sugar 'treats', an overabundance of commercial goods; alcohol, caffeine, nicotine, recreational drugs, hard drugs, pharmaceutical drugs, sex, porn, masturbation, food, and even music. All these things – because of an UNHEALTHY attachment to them via our emotional and psychological states – are training children to become addicts, and maintaining adults as addicts. And we are all responsible for influencing each other that this is acceptable, normal behaviour. It isn't.

The argument that all things are fine in moderation may be true of a lot of things, but not drugs, or anything that acts on you *as a drug would.* In fact, people who use drugs more infrequently are actually likely to be addicted to a greater extent than chronic addicts – on a psychological level. Yes. You read that right. Chronic addicts cannot help but be aware of the damage they are doing to themselves because it is extreme and not balanced by 'respectable' behaviour. They are more aware of the true nature and effects of the drug they are taking, and the inevitable repercussions for themselves and those around them – while, at the same time, being acutely aware of the hypocrisy of those around them (who would criticize or judge them because of their illness) who are taking the same drugs.

So-called casual or social users – which is most adults – are more easily able to fool themselves into thinking of their drug-taking as normal, acceptable, and even 'sophisticated'. Mostly, they are not even aware that they are psychologically and/or emotionally hooked. The hardest prison to escape from is the one you can't see – and the inmates will defend this prison and even work to stay locked up in it.

Let's take wine as an example. Wine drinking in particular is seen as adult, mature, sophisticated and socially acceptable. We talk about a wine's 'maturity' – the irony smarts like an upper-cut. There are even those who think of themselves as wine 'connoisseurs'. How ridiculous. Ever heard of a heroin connoisseur? It's just as absurd.

The only thing that a wine connoisseur is actually doing is differentiating between the additives in a given bottle or bottles, and noting the length of time and perhaps the method by which that wine has reached 'maturity' (again, what insufferable irony – especially since the drug Alcohol ACTIVELY PROMOTES immaturity, and a return to primal instincts that are ungoverned and not tempered by reason, which is why so much FUCKING and FIGHTING are done while drunk).

People drink wine because it has alcohol in it. The end. The fact that it is seen as a more sophisticated way to drink is utterly ludicrous, and is upheld only by its association with innumerable images and ideas that seem to support it.

The different additives, length of maturity, type of grape, etc., merely indicate how well the alcohol has been DISGUISED in the drink. Same goes for cocktails, punch, mixers, alco-pops – in fact, just about every drink. Only shots, shorts and spirits are closer to the truth of the drug – you know, those drinks that taste so fucking bad they make your face contort involuntarily.

A 'good' wine might give you less of a hangover, but again, the alcohol is still having the same effect. If it was really about the taste of the drink, why put alcohol in it at all? Or to put it another way, if we all woke up tomorrow and all the wines of the world had had their alcohol removed, which ones would people still want to drink because they taste nice? 'Oh, that's a lovely wine!' – what this comment really means is: 'The alcohol in this wine has been really well disguised!' It's got nothing to do with hints of cherry blossom or cheeky noses.

Alcohol is a drug, and people who sell it are drug-pushers (and people who drink it are drug-addicts). The fact that it's all dressed up in corporate regalia is irrelevant. There's no difference between wine 'merchants' and wine 'connoisseurs', and street-corner drug-pushers selling crack and their crack-addict customers.

Whilst chronic drug addicts make convenient scapegoats for the rest of society's drug-taking behaviour, we all have a personal and social responsibility to question, analyse, and understand the reasons for our own addicted patterns of behaviour.

Most people don't want to do this because it makes you a 'kill-joy'. The fact that taking drugs has got NOTHING whatsoever to do with genuine joy doesn't even cross most people's minds. They don't want their bubble to be burst.

The truth is that alcohol actively promotes immaturity, selfishness, irresponsibility (hence the idiocy of the 'Please drink responsibly' catchphrase – that's like saying 'Please snort cocaine responsibly'. Drug-taking and responsibility are mutually exclusive), narcissism, disconnection, and primal responses...as in pub brawls and one-night stands. There's absolutely nothing mature, sophisticated or acceptable about it.

That's the brainwashing, or SOCIAL CONDITIONING – commonly called SOCIAL ENGINEERING now – trying to stay alive in your mind because you see it as a part of yourself. Tradition. History. Remember, your brain has learned that to drink poison stimulates a big release of D+S, and it doesn't want that system dismantled because it has come to rely on it via your Ego and it's weaknesses and insecurities. Had a tough day? Have a drink. Feeling down? Have a drink. Feeling anxious? Have a drink. Going to a social function? Have a drink. Had a baby? Have a drink. Getting married? Have a drink. Meeting old friends? Get drunk. Economic woes? Have a drink. Relationship problems? Have a drink. Bored? Have a drink. Feeling inadequate? Have a drink. Depressed? Get drunk. It's your birthday? Get drunk. It's Friday night? Get drunk. On holiday? Have a drink. Have lots of drinks.

It's all biochemical really, but it informs your thought patterns, belief systems, and even feelings – in fact, mostly feelings. You take drugs to *change the way you feel (about yourself)*.

ONCE THE PRISON OF DRUG ADDICTION IS IN PLACE, THE INMATES WILL FIGHT TO KEEP IT IN PLACE.

That single prison of alcohol addiction in society at large corrupts so many other aspects of Humanity. People talk about 'the Human Condition'. I can sum it up in one word: ADDICTION. Within that one word, there is another invisible word: ATTACHMENT. Chronic addiction is really chronic attachment. Attachment to a way of life based on the brain's learning of how certain behaviours produce certain biochemical results. It's a formula. That's all.

This is one reason why Spiritual Masters practice the art of non-attachment. If you're not attached to anything on a psychological or emotional level, you cannot be manipulated. You cannot be bullied. You cannot be sold a lie as a truth. You have clear vision. You have a genuinely objective perspective.

So, you might ask, how come some people become chronic addicts, but most seem to be satisfied with a moderate amount, or seem to be able to 'take it or leave it'.

The answer is that most people are only experiencing a certain level of addiction to the drug, which as you will recall, is an addiction to the triggered release of D+S. They experience the effects, and have some amount of psychological or emotional attachment to the drug via social conditioning, but there is no NEED for them to do it. They have not developed a physical dependency. There are two main reasons for this: one is that there is a lack of motivation on a deeper level to develop a physical dependency (not much fear or anger, no particular trauma, etc.), and the second is that they are not constitutionally strong enough to be able to sustain a physical dependency.

Some people are more affected by the social conditioning, whereas chronic addicts are DRIVEN to seek the D+S release to satisfy a need on a deeper level. Addicts are also affected by the social conditioning factors, of course, but their need for D+S release is far greater than the average person's, and will continue to increase. Let's not forget that the more D+S that is artificially released, the more D+S receptors are shut down in the brain as it logically concludes that fewer are needed because the amounts of biochemical opiates being released is so great. In a vicious circle, you then need more D+S released for the few remaining receptors, which means you need more and more poison in order to secure this greater release.

An addict's need for comfort or relief is greater. This need is most often created by some trauma or other (which, by the way, explains how you can quickly develop a drug dependency at any time in your life). For example, in myself, I discovered that there were two powerful negative emotions that created in me the need – or desire – to drink alcohol much more than the average person. These emotions were Fear and Anger. So, essentially, you drink to cope with feelings you can't handle, or to 'escape' from them for a while. But as we all know, when you come back from the escape, your problems are waiting for you like an obedient hound – and now you have a hangover as well.

Chapter Three:

Temptation

I've covered Pull Factors in chapter one. But what are the Push factors? Why do we start taking drugs? We don't when we're children. Why do we start trying to solve problems or resolve feelings or 'enjoy' life by taking drugs? I've referred to it already.

Two words: Social Conditioning.

I don't mean just the wider society out there somewhere... I mean in your personal environment, and at a time (usually the teens) when you are trying to define your personality. I mean your home, your family, your friends, the TV in your home, the movies you grew up with, even the novels you read, pictures you looked at, posters you had on your walls, music you listened to, your extended family, cultural identity, your hometown. All these elements have an effect on you growing up and literally BRAINWASH you with what behaviour is acceptable, and what sort of behaviours you can expect to experience as an adult. More than that, I believe when you start taking drugs you traumatise the body and this has two effects: one, you start becoming a different version of yourself from this point on, and two – and most importantly – you 'freeze' a part of yourself in this moment which means every time from then on that you take the drugs, you remind yourself of the person you were just before you first started taking them. I believe this is a major cause of the anger and depression that lots of addicts feel. There is evidence to suggest that every time you traumatise the system your personality, as it is, STOPS – and you go off on a tangent. You spend the rest of your life trying to stop taking drugs so you can get back to your original personality, and continue. Smoking is the perfect example.

I digress. In a nutshell, an adult's behaviour is a walking advert for a child growing up, repeated over and over and over. Your Grandad sipping whisky and smoking a cigar, your parents getting drunk at Christmas or on holiday and trying to justify it by its seasonal context, your uncles and aunts drinking at every family reunion, and just about every adult you know drinking and/or smoking at EVERY social event from Weddings to Christenings to Funerals. Birthday parties, barbeques, Easter, Halloween, bank holidays, annual holidays, engagement parties, stag nights, hen nights, Friday nights, sporting events, and so on.

The calendar is peppered with excuses to drink and smoke, or take other drugs. And that's not including regular pub nights, dinner dates (where it's seen as adult sophistication), or drinking in the home as a matter of course. Many people have drinks cabinets in the home, and even if they do not partake, it's at the ready to offer to anyone who does.

How can a child growing up in this environment be expected NOT to take drugs, or at least try them? All the adults are doing it!! The psychological blueprint is set. By the time you get to your teens, you have been observing and absorbing this bullshit for well over a decade. Your brain is now hard-wired. If you're unlucky enough to have suffered some trauma early in life, or an unhealthy environment, which has created a predominant emotional state of Fear, Anger,

Depression or whatever, what are the chances that you WON'T develop an unhealthy attachment to hits of D+S (which your brain quickly learns makes you *feel* better) once you start trying/taking drugs?

Chapter Four:

Behind Closed Doors

Family factors, or rearing, is likely to have the most significant effect in determining what kind of a drug addict you turn out to be.

There, of course, other factors involved such as how strong your constitution is, how susceptible you are to the Social Conditioning and, as previously discussed, whether or not there are particular 'Push' factors – like Fear – in your psychological and emotional make-up.

Even though Push factors may come from something like being bullied at school, unrequited love, a death in the family, etc., most research I've looked at suggests that the majority of psychological and emotional difficulties are created in the home environment – usually unwittingly – during the period of child-rearing. So, 0-18. And more specifically, they are the result of the parent-child relationships.

In most cases, no blame can be assigned since most parents do not maliciously cause problems for their children (although I've known some people whose parents did do that, and that is not as uncommon a situation as we might think).

Your Child-Mother and Child-Father relationships are of the utmost importance. These will determine the balance and level of health of the Masculine and Feminine Energies WITHIN YOU.

They set the tone for all future relationships. Of course, you would have to have no negative behaviours at all to have no negative impact on your child, and your relationship as parents would have to be in perfect harmony. That is unrealistic since the behaviours and negativity are passed down the generations – it's a chain-reaction.

Plus, living as we do in a polar reality, negativity exists, so you have to be able to learn to cope with that: pain, discomfort, failure. What is most relevant though is the level of harmony or disharmony in the home. The level of balance. If negative behaviours are dominant, you're likely to have more to deal with in your own behaviour.

So, for example, people who drink who have been raised in a predominantly healthy, balanced atmosphere are far less likely to develop serious Push factors. They don't have enough need to escape that would lead them to a physical dependency. Therefore, they are far more likely to be able to take or leave drugs of any kind. On the other hand, let's say someone is raised in a home full of vitriol – perhaps physical or sexual abuse, verbal abuse, domination, manipulation, neglect, etc. – then it is more likely for that person to develop Fear, Anger, Depression, Hate, Resentment (a Victim Mentality), and therefore more of a need, desire, or tendency to seek large hits of D+S at a high level of frequency – as a COPING mechanism. Their need to 'escape' will be greater.

It is an escape because when your system is flooded with biochemical opiates you literally cannot feel any mental frustration or emotional pain for a while. However, once you pass the initial high, with alcohol in particular, your tolerance levels kick in and you are then able to feel the original mental frustration or emotional pain...***even though you are on drugs.*** This explains why chronic addicts will drink to the point of unconsciousness, or 'oblivion', as it is called. You're trying to drown out the pain, disappointment, anger, stress, frustration – or, you're trying to maintain the high. Same thing.

The nature of addiction is that as you use drugs to provoke bigger amounts of D+S in your system, your brain shuts down some of its own receptors for the D+S to attach to as a way of regulating things or because it doesn't need so many. This, in turn, prompts the addict – who had too few receptors in the first place! – to seek even bigger hits of D+S which eventually shuts down more receptors. It's a vicious cycle, an inevitably worsening downward spiral and that's why, over time, the addict will increase 3 things: the VOLUME of drugs consumed; the STRENGTH of the drugs consumed; and the FREQUENCY with which the drugs are consumed. This also explains why many addicts will combine activities which have some similar sort of effect. They may drink AND overeat. They may drink AND smoke. They may drink AND overeat AND smoke. Others will combine prescription drugs AND drink, or combine hard drugs AND prescription drugs. In the end, you can't get high enough, and this can lead to death. Also, if you are trying to quit and you have undergone a period of abstinence but then you 'slip' and return to the drug, you can easily overdo it and kill yourself that way. The addict within has been deprived, so if you return to the drug after a break, be it 6 months or 6 years, the addict within will make full use of the opportunity to 'catch up' and you are likely to over-medicate yourself – also, commonly, to the point of death (as was the case with Amy Winehouse, for example).

Finally, certainly with alcohol, your brain can become chemically altered to such an extent that you would not be described as 'being of sound mind' *even if you were sober.*

It's crucial to nurture in yourself – and your children AND your inner child – as many simple, healthy, balanced, grounded, soul-nourishing pursuits as possible.

Chapter Five:

The Power Of Association

The big Corporations that produce the socially acceptable drugs (that they want you to take) will use manipulation techniques – or 'brainwashing' – with great precision and aggression in order to get you to buy their product, and keep buying it. It's called 'stimulating the market'. They'll use any association that panders to your neuroses, or emotional or psychological inadequacies. Any association, in fact, that diverts your attention from the truth that they are selling drugs, and that they are DRUG PUSHERS. This, by the way, includes the pharmaceutical companies – or 'Big Pharma', as it's now called – who do produce necessary medicines but also encourage an unhealthy attachment to over-the-counter prescriptions by feeding on your FEARS.

No-one's going to say 'Here's a poisonous liquid. We call it alcohol. You should drink it.' Instead, you'll be presented with an *associated image* which depicts something you aspire to. The 'perfect' relationship; the high-life; the cool image; tradition; bohemian abandon; poetic wistfulness; urban toughness; business shrewdness; whatever. The Corporations sell you a vision – an illusion – and *associate* it with their product. 'This is what your life could be like…IF you consume our product.' After that, the association itself is enough to continue your slavery to the product. You feel that by consuming the product you have achieved some aspect of the illusion. You can observe this delusion being acted out in daily life if you look closely.

Another form of social conditioning, and probably more powerful and influential (as previously mentioned), is the influence of your immediate family, extended family and friends. We all influence each other in this regard by buying into the same bullshit. Each generation influences the new one growing up. Seeing your parents drink, smoke and drink caffeine and associate these drugs with every aspect of living from upset to celebration creates that same association in your own mind as being normal while you are still a child. By the time you become an 'adult', you are PRIMED FOR ADDICTION.

Then you have the conditioning of the drug itself. Take smoking for example. Most people who smoke do so on a frequent basis. This is because nicotine is a fast-acting drug, and leaves the system rapidly. As a result, you 'come down' off your hit before long and so you soon feel you need another. Each time you smoke, a subtle but powerful association is made between the smoking and whatever situation you happen to be in, or whatever you're doing at the time. Pretty soon, you associate smoking with everything! Nights in, nights out, sunny days, rainy days, cups of coffee, cups of tea, pints and shorts, being busy, being stressed, being bored, relaxing, dating, after sex, and even after exercise!! And that's another point: what magical drug can work the same in all these different and even opposing situations?! Certainly not nicotine! It APPEARS to work in all these situations because whatever you are doing, a release of Dopamine and/or Serotonin will always make you feel more comforted or better. It's not the activities that matter. It's getting to the next hit.

Unfortunately, with smoking in particular, because you end up associating it with EVERYTHING, this is one of the main reasons it's so difficult to stop. Everything reminds you of it!! EVERYTHING is a trigger to smoke!

The Power of Association should not be underestimated. By the time you've spent years of your life repeating these links between smoking and virtually everything else, it has become a *deeply ingrained pattern.* There is evidence to suggest that repeated thought processes (which, by the way, is how you derive a sense of your SELF), which usually have an association of some sort (even healthy ones), actually create grooves in the brain. The pattern of your brain and mind is being created and informed by drug use.

Another Social Conditioning fallacy is that alcohol relaxes you and relieves stress. It does not. It masks the stress and, because it's an anaesthetic, it forcibly relaxes the muscles, but that's not the same thing. True relaxation can only be achieved while sober, and involves a conscious awareness that you are taking the time out to relax, or that you are engaging in an activity that you find relaxing. Like going for a sauna, for instance. Or lying down. If you want to relax your mind and body, try meditation. Or sleeping.

Finally, the 'warming' effect of alcohol. When you ingest poison, your body panics because it is under attack. It sends your blood rushing to the vital organs to help protect them, and this rush of blood is perceived as a warm inner glow. At the same time it removes blood therefore from the body's extremities, exposing them to a greater vulnerability – especially to cold.

PART TWO:

RE-TRAIN YOUR BRAIN

Chapter Six:

Why The Drugs Won't Work

The drugs won't work because they've never worked. All they do is give you a false and temporary sense of comfort, relief or elation. It's not real. It's not real because it's artificially stimulated. Real and lasting senses of relief, comfort or elation come from healthy activities, and build an overall sense of self-esteem, confidence and inner peace.

The other main reason they won't work is because, over time, the highs become less and the lows become more exaggerated. It's a downhill run, all the way to six-feet-under.

Eventually, you get to the point where you want to be high all the time – chronic addiction – but even someone with massively increased tolerance levels can't sustain this indefinitely. In the end, the body just packs up.

If you're one of those people who can take it or leave it, you're living in a Fool's Paradise. You're living out the *illusion* of happiness, instead of giving yourself the opportunity of experiencing true happiness, and you'll always feel that life is somehow more exciting, enjoyable or bearable when you do take drugs, but you won't know why. Unless you're reading this – in which case, you already know why, or have just found out. It's an illusion. A false reality.

A lot of people don't know how to 'enjoy' themselves without alcohol, and that's genuinely sad. You should be able to go to any social occasion and give yourself the option NOT to drink alcohol, and STILL have an excellent time (and not because 'I can't. I'm driving.') You should have an even better time, in fact, because it's real.

As for the idea that drugs are an effective means of coping with life...er, no. When you get high, you may forget about your troubles briefly, but when you come down all the original problems are still there and now you have a hangover as well.

The only way to really deal with problems is to deal with them. Get therapy, read self-help books, go to anger management, do assertiveness training, take up a new hobby, build up your confidence, expand your social circle, eat the right food, rest properly, take some exercise, meditate, help in your local community, try yoga, do a new sport, read good books, paint, sing, create, whatever! But you will certainly not find the answers in any drug. Take it from me: they cause more problems than they *appear* to solve.

Chapter Seven:

Withdrawal And Relapse

Let's not beat around the bush, withdrawal can be a bitch. I went on an alcohol binge once and it took me a full week to sober up. The greater your tolerance level, the more drugs you can take – the longer it is before you fully recover.

There is a WARNING here though: be careful. Trying to quit cold turkey or come down too quickly can be extremely dangerous after years of abuse. I tried to go cold turkey off alcohol a few times and induced terrifying seizures that included a sense of panic and disorientation, increased heart rate, shakes, cold sweats, quickened breathing, pins-and-needles over the entire body. If you are a chronic addict, come down gently. You have to wean yourself off slowly, but don't use this as an excuse to continue either.

As previously discussed, other drugs are hard to kick because you associate them with every aspect of your life: mainly nicotine, but I'd add alcohol a very close second in this category. And bear in mind, *they're both legal and socially acceptable, and cause more social disruption and death than all the other drugs put together.*

The good news is that when you have recovered physically, all drugs are easier to leave behind...once you understand the mechanics of addiction. Biologically speaking, you need to stop provoking unnecessarily large amounts of D+S being released into your system and cause normal amounts through healthy activities and attachments, thereby giving your brain a chance to recover a normal amount of D+S receptors. This won't happen overnight, so it takes effort and you must be persistent. The exact opposite of the quick fixes of taking drugs.

Drugs remove the natural spiritual high from life while at the same time trying to replicate it. Or as an international drug trafficker I once knew put it: 'Taking drugs is basically a fucked up way of trying to feel closer to God.'

Drugs make normal life seem mundane, arduous and problematic. The truth is that drug-taking becomes mundane, arduous and problematic.

Chapter Eight:

Rehabilitation – Healthy Resources

You need to learn to live again. It's all very well putting the drugs down and coming to a point of acceptance about what's happened in your life, but you need to LEARN TO LIVE AGAIN. Yes. I repeated that on purpose. That's because it's worth mentioning twice. Worth mentioning twice.

I had to start completely from scratch at the age of 27, and it's taken years (15) to sort it all out. However, I went a very long and roundabout route, and I'm sure it can be done quicker and more efficiently. But certainly, one thing I learned is that you must be patient with yourself and allow time for readjustment and re-learning.

If you are a chronic addict, or a long-term addict, you must accept that you have been thinking, feeling and reacting as a drug addict for ages. You're unlikely to undo that overnight. But you can support your psychological and emotional changes by reading books about the mechanics of addiction (see Bibliography), other people's stories, etc., and by reading material that supports the kind of life you want to change to, plus support groups and the healthy activities already discussed.

It's also quite important to make your peace with the everyday activities that need to be done for the smooth and enjoyable running of your own life: cleaning, washing, cooking, ironing, vacuuming, etc. This will also ground you.

I always used to see these things as unexciting chores that I didn't FEEL like doing, but really, to take care of these things is taking care of yourself. It's saying: 'I love and support myself enough to ensure that I live in a clean and healthy environment.' Slightly different from the situation I was in as a chronic addict where the flat I lived in was full of dirty plates and mugs, empty food containers, cans and bottles, unwashed clothes for weeks on end, unchanged bedclothes for months at a time.

The other reason the everyday ordinary activities are so important is that they give you a reconnection to normal life patterns, and a sense of self-esteem; a sense of being in control of your everyday life; a focus, if there's nothing else you want to do. You have to nurture yourself and support your environment in the right way.

Chapter Nine:

The 'Rebel' Myth

There's a Myth...that if you take drugs, drink and smoke, that you are wild and exciting, a bad-boy, a Rebel. It's bollocks. If you think you're rebelling against society – as I did – you need to wake up. Society is an extension of your parents. If you think you're rebelling against your parents – as I did – you need to wake up. The only person you really hurt in this 'Act of Rebellion' is yourself.

It's pretending to do the opposite, but it's hurting YOU. It comes disguised as a fun, exciting Friend who'll be able to give you rushes of distraction and good feelings. Then – slowly, but most surely – it becomes your Enemy. It tries to beat you down from within. As your drug-addled Ego struggles to survive, it turns on itself.

Most people start experimenting with drugs in their teens or early twenties. Teens, in particular, is when you're most at risk because you're moving from being a child to being a young adult, and in that process, you're defining yourself. WARNING: DON'T DEFINE YOURSELF AS AN ADDICT.

If you're having a shit time in the home and you turn to drugs, what you're really doing is saying to your parents: 'These drugs can give me the love that you can't give me.' But there's a problem with that.

All negative behaviours are handed down generationally, unless they are addressed and dealt with. Your parents, and the rest of society, have their own problems that have been handed down to them. They're not perfect. However, if they dump this baggage on you, whether intentional or not, and whether it's neglect, abuse, Fear,

Anger, or whatever...you've got two choices, ultimately: One, you react – or 'Rebel' – against it, and continue living out that repeating pattern, or Two, realise that people who hurt you – intentional or not – are already hurting themselves, and support and love YOURSELF by choosing healthy pursuits.

In the end, you'll be stronger than anyone who didn't or couldn't love you properly – and you'll develop compassion for them in their weakness – because you will have dealt with the things they couldn't face about themselves. That's TRUE rebellion. Getting wasted on a Saturday night is a copout. An escape. A distraction. An illusion.

Sticking two fingers up at the rest of the world might make you feel better, and might help you to convince yourself that you can't be hurt, but you're already hurt or you wouldn't be doing it. When you take drugs or become an addict, really, you're sticking two fingers up at yourself. You think you're doing yourself a favour, but actually, you're doing the opposite.

If you really want to be a Rebel, change yourself.

Conclusion:
Spirituality

I have come to the conclusion that Love is the only Truth. Pretty much everything else is an illusion created by the Ego. The Ego does serve a genuine purpose of self-protection, but that has mutated into SELF-PROJECTION. If you don't love yourself properly – and you can't do it by taking drugs – you can't love your life properly, you can't love other people properly, you can't love Nature or the Earth properly. According to the Law of Attraction, the way you treat yourself will attract that treatment from others, but you will blame them for it. Example: I used to take no care or pride in my appearance. Some would observe: 'You look like a tramp.' Rude? Yes. Hurtful? Yes. BUT...if I was looking after myself properly and dressed and cleaned accordingly, would I have attracted such a comment?

You see, any kind of drug-taking behaviour is essentially a Spiritual sickness. We can explain it on physical, emotional, psychological and even biological and chemical levels. But there is a deeper level. The Ego will try to magnify itself in whichever guise you have originally defined it because it wants to live – it wants to continue to be made manifest.

However, when you are doing all the things of life for yourself in a right way, supporting and loving yourself, you raise your SPIRITUAL VIBRATION to a higher level.

That is Success, in every respect.

Bibliography

The Brain Book – Peter Russell

Unleash The Giant Within – Anthony Robbins

The Big Book – AA

The Easy Way To Stop Smoking – Allen Carr

The Only Way To Stop Smoking Permanently – Allen Carr

The Easy Way To Control Alcohol – Allen Carr

How To Stop Smoking And Stay Stopped For Good – Gillian Riley

They F*** You Up – Oliver James

Addicted – Tony Adams

Addict – Stephen Smith

Blessed – George Best

In The Realm Of Hungry Ghosts – Dr. Gabor Mate

The Hidden Messages In Water – Dr. Masaru Emoto

The Power Of Now – Eckhart Tolle

How To Be Your Own Best Friend – Mildred Newman/Bernard Berkowitz/Jean Owen

The Middle Passage – James Hollis

The Biggest Secret – David Icke

Iron John – Robert Bly

Body Language – Allan Pease

How To Stand Up For Yourself – Dr. Paul Hauck

I'm OK, You're OK – Thomas Anthony Harris

Leaving Las Vegas – John O'Brien

Evil Spirits: The Life Of Oliver Reed – Cliff Goodwin

Instant Confidence – Paul McKenna

Change Your Life In 7 Days – Paul Mckenna

On Learning From The Patient – Patrick Casement

Bare Knuckle Fighter – Bartley Gorman

From Gangland To Promised Land – John Pridmore

The Spiritual Teachings Of Seneca – Mark Forstater

The Spiritual Teachings Of Marcus Aurelius – Mark Forstater

The Spiritual Teachings Of Swami Premananda: Volume Three – Sri Premananda Trust

Autobiography Of A Saint: St. Therese Of Lisieux – Ronald Knox

Autobiography Of A Yogi – Paramahansa Yogananda

Assertiveness At Work – Ken Back/Kate Back

Emotional Intelligence – Daniel Goleman

The Problems Of Philosophy – Bertrand Russell

www.ingramcontent.com/pod-product-compliance
Lightning Source LLC
Chambersburg PA
LVW081051170526
758CB00006B/1935